Industrial Plastics Fire:
Major Triage Operation
Flint Township, Michigan

Investigated by: Tom D. Copeland

This is Report 025 of the Major Fires Investigation Project conducted by TriData Corporation under contract EMW-88-C-2649 to the United States Fire Administration, Federal Emergency Management Agency.

Homeland
Security

Department of Homeland Security
United States Fire Administration
National Fire Data Center

U.S. Fire Administration Fire Investigations Program

The U.S. Fire Administration develops reports on selected major fires throughout the country. The fires usually involve multiple deaths or a large loss of property. But the primary criterion for deciding to do a report is whether it will result in significant "lessons learned." In some cases these lessons bring to light new knowledge about fire--the effect of building construction or contents, human behavior in fire, etc. In other cases, the lessons are not new but are serious enough to highlight once again, with yet another fire tragedy report. In some cases, special reports are developed to discuss events, drills, or new technologies which are of interest to the fire service.

The reports are sent to fire magazines and are distributed at National and Regional fire meetings. The International Association of Fire Chiefs assists the USFA in disseminating the findings throughout the fire service. On a continuing basis the reports are available on request from the USFA; announcements of their availability are published widely in fire journals and newsletters.

This body of work provides detailed information on the nature of the fire problem for policymakers who must decide on allocations of resources between fire and other pressing problems, and within the fire service to improve codes and code enforcement, training, public fire education, building technology, and other related areas.

The Fire Administration, which has no regulatory authority, sends an experienced fire investigator into a community after a major incident only after having conferred with the local fire authorities to insure that the assistance and presence of the USFA would be supportive and would in no way interfere with any review of the incident they are themselves conducting. The intent is not to arrive during the event or even immediately after, but rather after the dust settles, so that a complete and objective review of all the important aspects of the incident can be made. Local authorities review the USFA's report while it is in draft. The USFA investigator or team is available to local authorities should they wish to request technical assistance for their own investigation.

This report and its recommendations were developed by USFA staff and by TriData Corporation, Arlington, Virginia, its staff and consultants, who are under contract to assist the USFA in carrying out the Fire Reports Program.

The USFA appreciates the cooperation received from the Flint Township, Michigan, Fire Department. Particular thanks go to Fire Chief Donald S. Rowley and Captain Robert W. Hamlin.

For additional copies of this report write to the U.S. Fire Administration, 16825 South Seton Avenue, Emmitsburg, Maryland 21727. The report is available on the Administration's Web site at http://www.usfa.dhs.gov/

U.S. Fire Administration
Mission Statement

As an entity of the Department of Homeland Security, the mission of the USFA is to reduce life and economic losses due to fire and related emergencies, through leadership, advocacy, coordination, and support. We serve the Nation independently, in coordination with other Federal agencies, and in partnership with fire protection and emergency service communities. With a commitment to excellence, we provide public education, training, technology, and data initiatives.

TABLE OF CONTENTS

Industrial Plastics Fire: Major Triage Operation
Flint Township, Michigan
November 1988

Local Contacts: Donald S. Rowley
Fire Chief
Flint Township Fire Department
6-5331 Reuben Street
Flint, Michigan 48532
(313) 732-4413

Robert W. Hamlin
Captain
Flint Township Fire Department
G-5331 Reuben Street
Flint, Michigan 48532
(313) 732-4413

OVERVIEW

On Tuesday, November 29, 1988, a fire occurred in a one-story Flint Township, Michigan, factory that made polyurethane automobile bumpers. Chemicals used in manufacturing the bumpers and a large quantity of the bumpers burned. The fire resulted in about 96 people being checked at the hospital and the destruction of part of the metal building housing the factory, at a loss of 3,000,000 dollars. The primary problem in fighting this fire was toxic smoke and the problem of having to access the fire from downwind. A county "fire coordinators" system of Incident Command was successfully used to coordinate the activities at the fire and an evacuation of 150 people.

THE FIRE DEPARTMENT

Flint Township is located in Genesee County west of the city of Flint, Michigan. It has a population of about 37,000 people and covers an area of 25 square miles. The Flint Township Fire Department has three stations with 12 career firefighters and about 44 on-call firefighters and officers. One career firefighter is on-duty at all times at each of the three stations plus the chief and assistant chief during the day.

1

SUMMARY OF KEY ISSUES

Issues	Comments
Cause	Unknown at the time of this report. Origin in 160 feet by 125 feet addition to rear of building. (See Appendix A for Site Plan.)
Detection	Discovered by passerby while building was unoccupied. Alarm was transmitted by telephone.
Building Structure	One-story, non-combustible/metal building. About 20,000 square feet fire area.
Contents	Fire area contained polyurethane automobile bumpers and associated chemicals used in manufacturing.
Fire Protection Equipment	No sprinkler system. Building Officials and Code Administrators (BOCA) Building Code would require sprinkler system and subdivision of building if built today.
	No detection or alarm system other than a security system.
Incident Command	County "fire coordinators" system successfully used during incident.
Injuries	Firefighting hampered by smoke from burning plastics. Only access was from downwind side. One hundred fifty five people triaged including one police officer and six civilians.
	Ninety-six people checked at hospital.
	Twelve people admitted to hospital, most for toxic smoke inhalation, plus one firefighter with broken shoulder.
Evacuation	About 150 people actually evacuated. News media erroneously stated as many as 3,000.

The Flint Township Fire Department equipment includes seven engines, two 100-foot ladder trucks, two van-type support units for air and lighting, and one van-type unit that is used for air. Four of the engines are typical pumpers. The other three engines carry 2,000 feet of 4-inch diameter hose.

The department has a formal mutual aid agreement with other fire departments in Genesee County, including the city of Flint. Mutual aid response is not automatic but results from a specific request, as typical elsewhere. Radio dispatching can be done from any of the township stations. Dispatching is normally done from Station #1, but all stations have the capability. If the station dispatching has to respond, dispatching transfers to the next available station until relief personnel arrive.

Genesee Central Dispatch is operated by the State from the highway patrol station. They dispatch most of the county fire departments other than the city of Flint and the Flint Township Fire Department.

The Flint Township Fire Department has no full-time inspector or fire investigator. These functions are conducted by other on-duty personnel as time permits. The Diverse Plastics, Inc., facility had not been inspected prior to this fire, nor had a pre-fire plan been prepared.

The Flint Township Fire Department has conducted and is continuing to update a survey based on the Michigan Right-To-Know Law concerning hazardous chemicals. A questionnaire is sent to businesses with a request that they complete the forms and provide certain information (Appendix E). The response to this has been good. However, the information on file for Diverse Plastics was not up-to-date.

FIRE COORDINATORS

A system has been developed within the county whereby "fire coordinators" are used during major incidents. Their purpose is to coordinate the use and movement of equipment, personnel, and other resources and to maintain fire protection throughout the county. The coordinators are specially designated personnel from the various fire departments in the county. The system is activated by the officer in charge at an incident by contacting Flint Township Fire Department Station No. 1. which is the operations center for the Genesee county fire coordinators. The coordinators are then contacted by radio or beeper. Coordinators are assigned to duties at the station and at the scene.

When the fire coordinators are mobilized, some typically serve as sector commanders at the scene, and others at the fire station where they make sure that the various county fire stations are covered. They coordinate staging of firefighters and equipment, oversee personnel relief, and provide other management functions. (For further details see Appendix F.) The system is activated about once a month. It was started about ten years ago after a large fire indicated the need for such a coordination system and to insure that stations are covered during a major fire.

The Flint Township Fire Department Station No. 1 was used as the coordination center, the staging area, and the triage center in this incident. The Chief of the Flint Township Fire Department had a command post at the scene along with sector commanders at the scene. No units responded directly to the scene on their own.

One hundred eight (108) self-contained breathing units were utilized at the scene. Many of the tanks were filled several times. About 13 fire coordinators were used. About 50 firefighters were on the scene at any one time. A total of 194 fire personnel from 27 departments participated.

THE BUILDING

The Diverse Plastics, Inc., facility was a one-story metal building with a small two-story office area on one end. The original building was 80 foot by 225 foot with the long dimension perpendicular to the main road. At the rear, a 166 foot by 125 foot additional had been constructed. (See Appendices A and B for additional details.)

Except for the office area, the construction utilized steel structural members and steel exterior wall and roof material. The interior typically had fiberglass batt-type insulation attached to the under-side of the roof and the exterior walls. The addition was separated from the original building by this same type of steel construction but had a 12 foot by 12 foot unprotected opening between the new and existing construction. The wall separating the two buildings was not fire-rated. The new addition and the existing building each were essentially undivided. The building had no automatic sprinkler system and no fire alarm system. There was a security alarm system.

A narrow driveway extended around the perimeter of the building. The main road (Lennon Road) ran east/west on the north side of the building. Beyond the paved drive, there were open fields and trees on each side. There were no nearby exposures.

The original building contained manufacturing equipment, plastic bumpers, and raw materials. The additional (fire area) contained finished plastic bumpers, chemicals in 55-gallon drums used to make the bumpers and some manufacturing operations. Additional raw materials in 55-gallon drums were located on the south exterior of the fire area. The bumpers were stored primarily in stacked metal racks.

The right-to-know information on file at the fire station for this building indicated that isocyanate in 55-gallon drums was stored in the southwest corner of the original building. However, this information listed the building under a different corporate name and was out of date.

Under the BOCA Building Code, a new building of this construction type, occupancy, use, and undivided floor area would both require automatic sprinkler protection and would exceed the undivided area limitations. The 1987 editions of both BOCA Building and Fire Prevention Code were in effect in Flint Township.

THE FIRE

On the evening of Tuesday, November 29, 1988, a retired firefighter passing the Diverse Plastics, Inc., building noticed smoke coming from the building. He determined that there was a fire and notified the Genesee Central Dispatch Office. At 1944, Genesee Central Dispatch gave the alarm to Swartz Creek Fire Department as a fire on the roof. Flint Township Fire Department overheard the alarm and advised Genesee Central Dispatch that the building was in their district. Flint Township Fire Department immediately dispatched an aerial and two pumpers, one of which was equipped with 4-inch hose. Flint Township already had equipment at the scene of a mobile home fire in another area. Swartz Creek Fire Department did not continue to the scene at this time.

No one was reported to have been in the building at the time of the fire. The company monitoring the security system for the building reported receiving an alarm sometime after the fire department received the alarm from the passerby.

At 1949, the first arriving fire apparatus entered on the east side of the property and found smoke coming from the addition. Access was gained into the addition at its southeast corner. A 2-1/2-inch handline was used for interior attack, and it was thought that the fire was of limited magnitude at that time.

The interior attack was not effective and the fire intensified. After a few minutes, the firefighters inside were forced out of the building. Units arriving on the west side of the building observed high heat conditions and discoloration of the exterior wall. Flashover reportedly occurred. Mutual aid and additional equipment were requested. At 1954, additional equipment started being dispatched. At 2002 and 2003, the north and south coordinators were alerted to respond (and implement their fire coordinator system).

A concern about the toxicity of the smoke arose very early in the fire. Staffing power was intentionally kept to a minimum at the site and virtually all active firefighters at the site used self-contained breathing apparatus (SCBA). At 2040, it was reported that an 18 mph wind was blowing toward the north.

Access to the property was limited to the perimeter driveway from Lennon Road on the north side of the building. Smoke from the fire was very low to the ground and affected access by this driveway. It was necessary for firefighters to walk through the smoke in order to obtain refilled air tanks.

After having to retreat from inside the burning addition, firefighting efforts were directed at protecting the 55-gallon drums on the north side of the building and preventing spread into the original (older) building. A master stream device was located in the original building and aimed southward through the opening in the wall between the original building and the addition. The device had a 2-inch solid stream tip with an estimated flow of 1,200-1,300 gpm. At times, the stream was directed across the separation wall and northward in the original building to cool the ceiling of the original building.

Two 4-inch supply lines were connected to the fire hydrant near the front of the building and one to the next hydrant to the east. The two lines from the front hydrant supplied operations on the east side of the building, and the third 4-inch line served operations on the west side along with tanker supply. Water supply was generally adequate. A 16-inch water main served the hydrant directly across the street from the factory.

The fire was prevented from spreading significantly into the original building. Later in the incident, when it was feared that the wind might shift and affect other populated areas, a stronger effort was made to extinguish the fire in the addition. The fire was under control about midnight, over four hours after being detected.

It is believed that the main contribution to the fire was the finished polyurethane bumpers and packing materials. The bumpers were stored in steel racks stacked vertically in the addition. Even though many of the drums were later discovered with bulges, the materials in the drums within the building may not have been significantly threatened due to the venting of the fire through the roof and the operation of hose streams. The lack of major structural deformation other than in the southeast corner of the addition, which collapsed, supports this theory.

DAMAGE

All of the contents of the addition were either consumed or heavily damaged by the fire except for materials that were recovered from the 55-gallon drums. The original building suffered heat and flame damage primarily in the ceiling area and near the addition. The remainder of the building suffered considerable smoke damage.

ENVIRONMENTAL MONITORING

No samples of smoke were taken during the incident. Water runoff was tested after the incident and was reported to be "clean." A professional cleanup company was used to recover runoff and secure the remaining chemicals. Forty-eight drums of materials were removed after the fire.

EVACUATION

The Flint Township Fire Department dispatcher advised the scene at 2009 about the information in the right-to-know response from the company. At 2019, a "key holder" for the facility was sent to the scene. At 2041, about an hour after the fire was first reported, the Flint Township Fire Department chief requested that an area one-half mile north and one-half mile west of the fire be evacuated. The Flint Township Fire Department dispatcher correctly understood the chief's request for evacuation of the quadrant from one-half mile north to one-half mile west of this location, and relayed this information to Genesee Central Dispatch. The evacuation was conducted by the police department. About 150 people were evacuated. Somehow, there were reports that the evacuation area was to be one mile in all directions. The media exaggerated the situation and reported that about 3,000 people were evacuated. How the area to be evacuated and the number of people evacuated became exaggerated is not known.

The television media reported during the incident that the chemical involved was polyvinyl chloride. It is not known how the media reached that conclusion. Reports of cyanide involvement also were issued during the incident. These comments developed outside the fire department.

Amateur radio operators are part of the fire coordinators system and have a fixed installation at Station 1. A total of about 30 members of the amateur radio organization (ARES) were utilized during the incident at the coordination center, at the scene, at hospitals, and other locations.

When the evacuation order was given, the Flint Township supervisor activated the Flint Township Disaster Plan (Appendix G). The Genesee County civil defense director also responded. Although the county disaster plan was not formally enacted, several elements were used to assist with coordination of the evacuation, public information, and communications with other agencies such as the county health department. The Red Cross operated evacuation centers. An evaluation of the evacuation was being undertaken by the civil defense director at the time of this report.

TRIAGE

Because of the toxic smoke conditions at the fire scene, virtually all personnel who operated at the scene were returned at some point to Station No. 1, where they were triaged. Genesee County paramedics were used under mutual aid for much of the triage operation. About 96 people were transported to hospitals via ambulances for further evaluation or treatment. Several ambulances were used to transport as many as four patients at a time. The transport operations ran almost continually throughout the incident. Twelve people were admitted, but most were released the following morning. Most people transported were treated for toxic smoke inhalation except for one firefighter with a broken shoulder. Seven civilians from the evacuation area were also treated and released. The high number of people triaged and treated is not so much a reflection of this fire being unusually severe or toxic as that the incident was well managed and appropriate precautions were taken.

Often firefighters and emergency personnel are not routinely checked by medical personnel after they have been exposed to smoke. It is quite possible that very few of the people sent to the hospital after this fire would have been evaluated or treated through their own initiative. Management of this incident through the fire coordinator system included management of the returning personnel as well as staging and other firefighting operations.

LESSONS LEARNED

1. **Using fire coordinators as part of the Incident Command System was effective.**

    This type of operation is most applicable to coordination of multiple small fire departments as opposed to operations within a single large city department. The coordinators are responsible for and manage much more than just fire scene operations. The fact that most of the coordinators are located at the fire station rather than near the fire scene allows them to be much more objective in their work and not distracted by the emergency itself.

2. **Combining operations such as triage, staging, decontamination, and rest and relief at one location proved to be a problem.**

    Accounts of the fire relayed by firefighters leaving the scene combined with exaggerated media reports contributed to anxiety among firefighters waiting to be assigned. There was an inadequate number of public telephones for emergency personnel to communicate with their families concerning their whereabouts and safety. It may be advantageous to have all of these operations separated from the actual command center.

3. **Departments need to have established procedures for gaining safe access to the scene when there is low-lying smoke.**

   The size of this fire was not particularly unusual for an industrial fire. The primary problem other than toxicity of the smoke centered on access. Access to the fire was primarily through the smoke. Relief personnel, fresh air tanks, equipment, and vehicles also had to be moved through the smoke. Where this cannot be avoided, procedures need to be developed so that this can be accomplished in the safest manner possible. People moving through such smoke, even outside, should use SCBA.

4. **Accurate pre-fire surveys of hazardous materials (Hazmat) present are needed so that emergency personnel can know what to expect.**

   In this case, there was no up-to-date information concerning Hazmat in the building addition where the fire occurred. In this fire the lack of information added to the apprehension experienced by the firefighters and to the exaggerated media reports.

5. **The lack of automatic sprinkler protection and subdivision of fire areas are believed to be the major contributors to the consequences of this incident.**

   Complete and appropriate automatic sprinkler protection most probably would have suppressed or contained this fire. Manual firefighting efforts in situations such as this cannot be expected to prevent major fire development.

6. **Evacuation instructions need to be clearly understood.**

   The evacuation in this incident appears to have been reasonably effective. But there was a potential for confusion when identifying the area to be evacuated. In this fire, the request that an area one-half mile north and one-half mile west be evacuated was somewhat ambiguous. A more standard terminology should be developed to define an area to be evacuated, particularly when face-to-face contact is not possible. Maps with pre-marked codes can be referred to more easily.

7. **A fire coordinator or other responsible person needs to be assigned responsibility for working with the media.**

   During this incident, incorrect information about the materials involved in the fire and the magnitude of the evacuation were publicly broadcast. A media coordinator can provide the proper names of any materials involved, the toxicity, and any precautions that are required.

8. **Additional frequencies and dispatchers may be necessary to handle the communications load in an incident such as this.**

   During this incident, the fire department communications dispatcher found it extremely difficult to handle all of the communications needs. Procedures are needed so that additional personnel can assist with dispatching to allow full attention to radio traffic as well as accomplishing the many tasks associated with an incident of this type. This is not always going to be possible and is an area that needs further study.

9. **With the growing problem of Hazmat and toxicity, along with the increased concern for firefighter health and safety, it may become standard practice for emergency personnel to be medically evaluated after a fire.**

Attention was drawn to this fire by the reports of so many people receiving medical attention and being sent to the hospital – not because of the magnitude of the fire or the number of people evacuated. Historically, this has been done primarily only when personnel complained or had obvious injuries. The numbers of people medically evaluated in this incident may have actually been a result of good management rather than the seriousness of the incident. In the past, many of the emergency personnel may simply have gone home with a "bad headache."

APPENDICES

A. Site Plan

B. Fireground Operations Diagram

C. List of Slides, Selected Photos, and Site Plan Showing Locations From Which Slides Were Taken

D. Material Safety Data Sheets

E. Diverse Plastic's Hazardous Chemical Survey Form

F. Genesee County Fire Coordinators Standard Operating Procedures

G. Flint Township Basic Disaster Plan

APPENDIX A

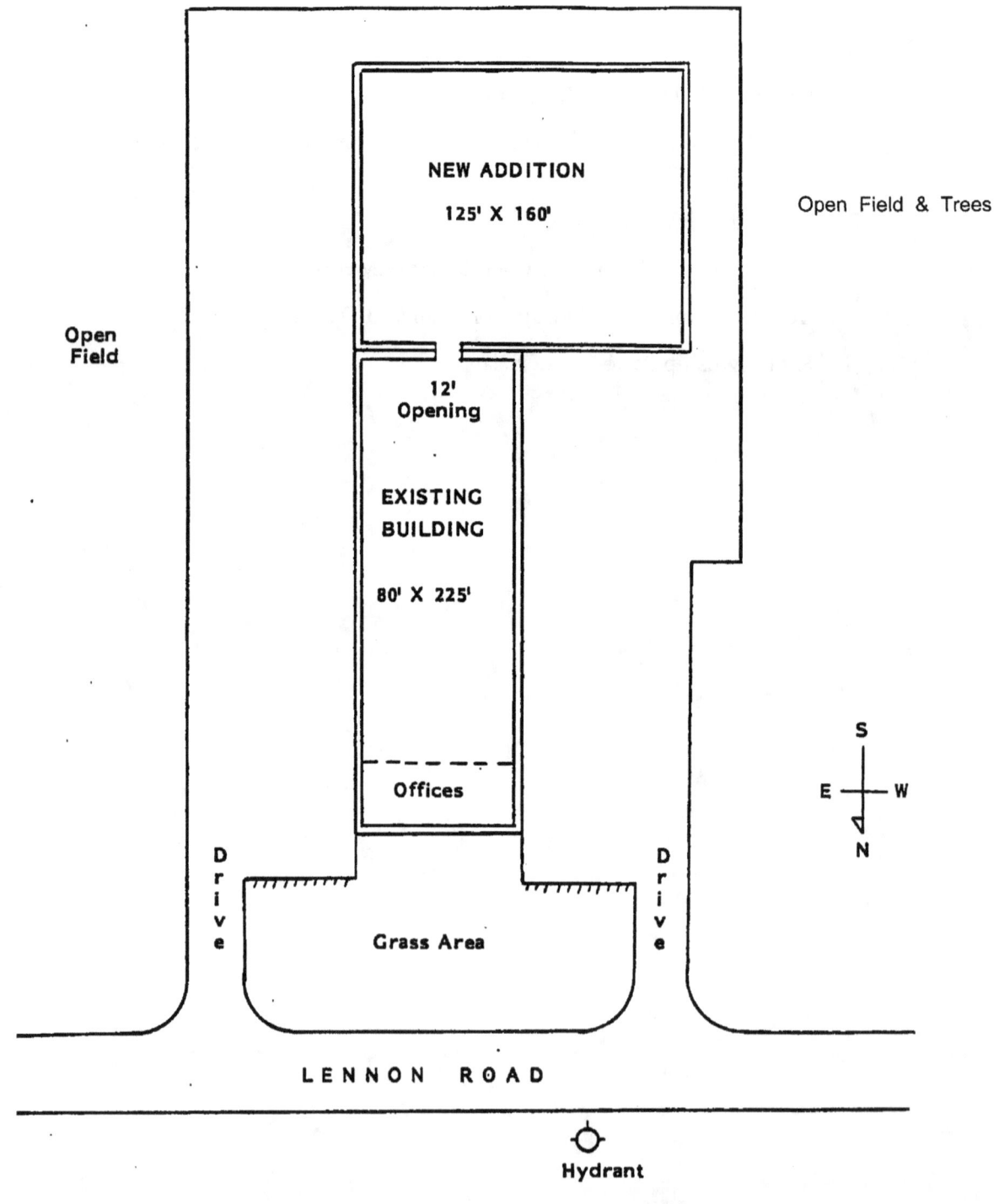

Open Field & Trees

NEW ADDITION

125' X 160'

Open
Field

**12'
Opening**

**EXISTING
BUILDING**

80' X 225'

Offices

S

E — W

N

D
r
i
v
e

D
r
i
v
e

Grass Area

L E N N O N   R O A D

Hydrant

267T1127	2-24-89
Diverse Plastice, Inc.	
NTS	202

APPENDIX B

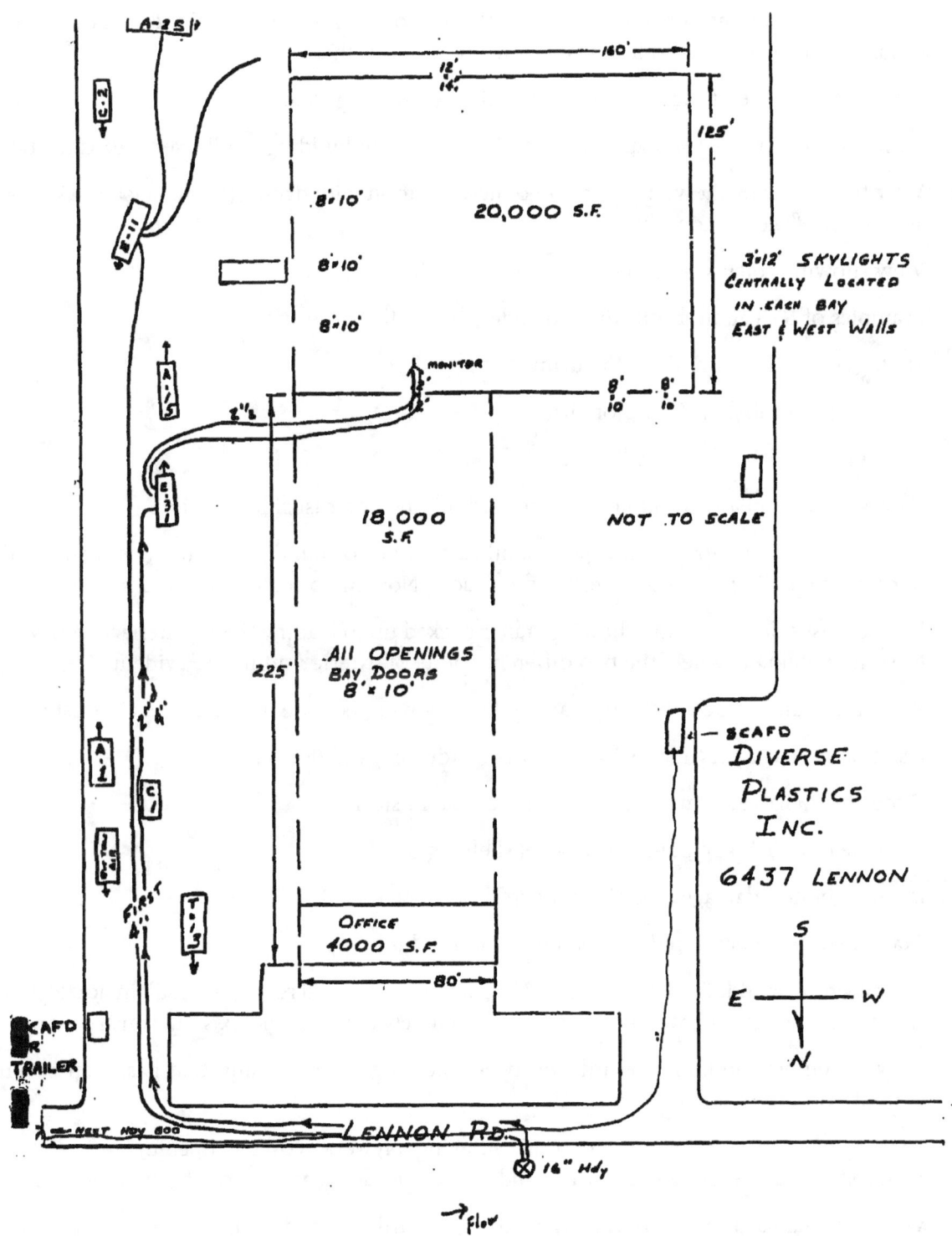

APPENDIX C

List of Slides

The slides with an asterisk have been made into photos and reproduced on the following pages.

1. Overview of building looking toward southeast from Lennon Road. Office area is in foreground to the left. New addition is in background to right.

2. View of north side of the new addition. Existing building to left.

3. View toward northeast showing south and west sides of building. Collapsed area is to right.

4. View from exterior showing south end of new addition with drum storage outside. Collapsed area is out of picture to right.

5. View showing collapsed southeast corner of building.

*6. Examples of 55-gallon drums that expanded due to fire exposure.

7. Photograph of label on 55-gallon drum.

8. Photograph of label on 55-gallon drum.

9. Photograph of label on 55-gallon drum.

10. View showing collapsed southeast corner of building and east side of building.

*11. View toward west showing collapsed southeast corner of building. Entry was made in this portion of building for initial interior fire attack. Note discoloration to building exterior.

12. View of east side of building showing trailer backed up to loading dock. Interface of new and existing building is where the two different colors of exterior siding are evident.

13. View into trailer backed up to the loading dock showing polyurethane bumpers in steel racks.

*14. Example of steel storage racks for bumpers inside new addition.

15. Close-up of charred remains of bumper material in steel rack.

16. Example of remains of bumpers on steel shelving.

17. Example of heat damaged plastic bumpers in steel racks.

18. Example of bumpers stored in existing portion of building.

19. View inside new addition toward north. Note steel columns and roof system still intact. Plywood area covers original unprotected opening between new addition and existing construction.

20. View inside new addition toward northwest showing door opening that has been covered with plywood.

21. View in existing building toward south showing plywood covered opening into addition. Monitor nozzle was located in this area and aimed into the new addition through the doorway.

22. View of metal partition from existing building toward new addition.

*23. Photograph taken at Flint Township Fire Department Station 1. Dispatcher room showing map and fire department information.

6. Examples of 55-gallon drums that expanded due to fire exposure.

11. View toward west showing collapsed southeast corner of building.
Entry was made in this portion of building for initial
interior fire attack. Note discoloration to building exterior.

14. Example of steel storage racks for bumpers inside new addition.

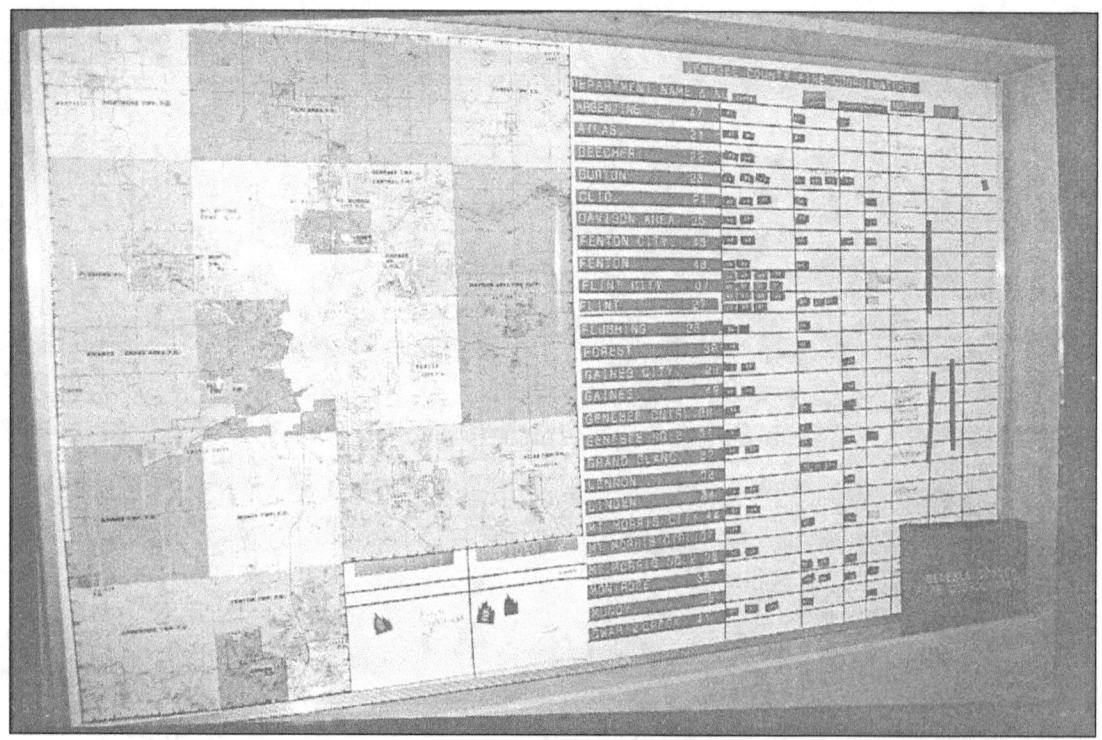

23. Photograph taken at Flint Township Fire Department Station 1. Dispatcher room showing map and fire department information.

Appendix C (Continued)

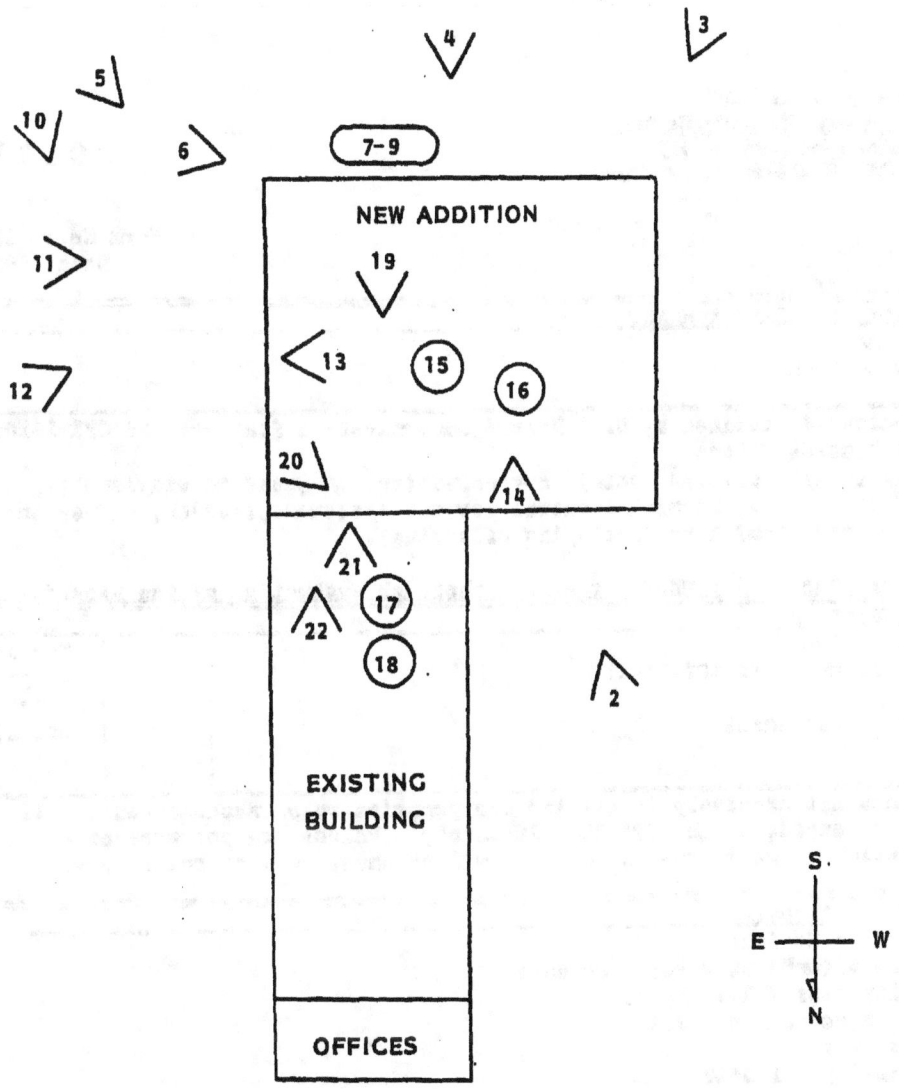

NEW ADDITION

7-9

EXISTING
BUILDING

OFFICES

Area or Direction of View of Slides
(Description Attached)

23 - Taken at Station One

267T1127	2-24-89
Diverse Plastics, Inc.	
NTS	267

APPENDIX D

Rubicon Chemicals Inc.
Wilmington, Delaware 19897
Phone (302) 575-5000 (24 Hours)

003213

Form No.: 6229
Date: 03/02/87

SECTION 1   NAME & HAZARD SUMMARY

Material name:
  RIMline® E-2150B

Hazard summary (as defined by OSHA Hazard Communication Standard, 29 CFR 1910.1200):
  Physical hazards: None
  Health hazards: Irritant (skin), corrosive (eye). Based on ethylene glycol -
    inhalation (TLV), harmful (central nervous system depression, kidney and brain
    injury, and harmful to developing offspring).

  Read the entire MSDS for a more thorough evaluation of the hazards.

SECTION 2   INGREDIENTS

	%	TLV (ACGIH)
Polyether polyol		Not listed
Ethylene glycol (CAS 107-21-1)		50 ppm, ceiling
Polyoxyalkylene amine		Not listed

Ingredients not precisely identified are proprietary or nonhazardous. All
ingredients appear on the EPA TSCA Inventory. Values are not product
specifications. gt = greater than, lt = less than, ca = approximately

SECTION 3   PHYSICAL DATA

Boiling point: No data
Vapor pressure (mmHg at 20°C): No data
Vapor density (air = 1): No data
Solubility in water: No data
pH: Greater than 8
Specific gravity: 1.0417
% Volatile by volume: No data
Appearance and odor: Liquid

SECTION 4   FIRE AND EXPLOSION HAZARD DATA

Flash point (and method): Above 230°F, 110°C (Setaflash CC)
Autoignition temp.: No data
Flammable limits (STP): No data

Extinguishing media:
  Water fog, foam, carbon dioxide, dry chemical, halon 1211.

Special fire fighting protective equipment:
  Self-contained breathing apparatus with full facepiece and protective clothing.

Unusual fire and explosion hazards:
  When misted in air, ethylene glycol becomes a moderate fire and explosion hazard.

16

Appendix D (Continued)

MATERIAL SAFETY DATA SHEET (continued)   RIMline® E-2150B

SECTION 5   REACTIVITY DATA

Stability:
Stable under normal conditions.

Incompatibility (materials to avoid):
Oxidizing agents.

Hazardous decomposition products:
Combustion products: Carbon dioxide, carbon monoxide, nitrogen oxides, ammonia.

Hazardous polymerization:
Will not occur.

SECTION 6   HEALTH HAZARD ASSESSMENT

General:
No toxicity information is available on this specific preparation; this health hazard
assessment is based on information that is available on its components.

Ingestion:
The acute oral LD_{50} in rat is probably above 5 g/kg. Relative to other materials, a
single dose of this product is practically nontoxic by ingestion. Acute ingestion of
ethylene glycol, a component of this material, can result in central nervous system
depression and kidney injury which can be fatal. The lethal dose for ethylene glycol
in man is estimated to be 1.4 ml/kg.

Eye contact:
This material can induce chemical burns if contact is made with human eyes.

Skin contact:
Short contact periods with human skin are not usually associated with skin irritation;
repeated and/or prolonged contact can induce skin irritation.

Skin absorption:
Ethylene glycol can be absorbed through the skin in toxic amounts.

Inhalation:
Vapors and aerosols can irritate eyes, nose and respiratory passages. Vapor
inhalation hazard of ethylene glycol is low because of low vapor pressure.

Other effects of overexposure:
Overexposures to vapors of ethylene glycol are associated with injuries to kidneys,
liver, lungs, blood and central nervous system. Skin absorption can induce central
nervous system depression and kidney injury. In addition to the effects observed
following acute ingestion of ethylene glycol, repeated ingestion can induce brain
injury.

Recently, results from two studies sponsored by the National Toxicology Program
(NTP) became available. In one study ethylene glycol was administered to pregnant
rats and mice by ingestion as large single daily doses. In the other study, ethylene
glycol was administered by ingestion as high concentrations in drinking water. Both
studies showed birth defects and deaths in some of the unborn animals. In earlier NTP
studies, no significant adverse effects were produced on either the reproductive
system or reproductive performance.

Appendix D (Continued)

MATERIAL SAFETY DATA SHEET (continued)   RIMline® E-21502

SECTION 6 HEALTH HAZARD ASSESSMENT (continued)

First aid procedures:

Skin: Wash material off the skin with plenty of soap and water. If redness, itching or a burning sensation develops, get medical attention. Wash contaminated clothing and decontaminate footwear before reuse.

Eyes: Immediately flush with plenty of water. After initial flushing, remove any contact lenses and continue flushing for at least 15 minutes. If redness, itching or a burning sensation develops, have eyes examined and treated by medical personnel.

Ingestion: Give 1 or 2 glasses of water to drink and induce vomiting by sticking finger down throat. Repeat until vomitus is clear. Refer victim to medical personnel. (Never give anything by mouth to an unconscious person.)

Inhalation: Remove victim to fresh air. If not breathing, give artificial respiration, preferably mouth-to-mouth. If breathing is labored, give oxygen. Consult medical personnel.

Note to Physician: Treatment for clinical cases of ethylene glycol poisoning can include ethanol infusions, which is based upon substrate competition for alcohol dehydrogenase. It has been proposed that renal toxicity is based partially upon the effects of metabolites produced by the action of alcohol dehydrogenase on ethylene glycol.

SECTION 7 SPILL OR LEAK PROCEDURES

Steps to be taken in case material is released or spilled:

Wear skin, eye and respiratory protection during cleanup. Soak up liquid with absorbent and shovel into waste container. Cover container and remove from work area. Wash residue from spill area with water and flush to a sewer serviced by a wastewater treatment facility.

Disposal method:

Discarded product is not a hazardous waste under RCRA, 40 CFR 261.

Container disposal:

Empty container retains product residue. Observe all hazard precautions. Do not distribute, make available, furnish or reuse empty container except for storage and shipment of original product. Remove all hazardous product residue and puncture or otherwise destroy empty container before disposal.

SECTION 8 SPECIAL PROTECTION INFORMATION

TLV or suggested control value:

No ACGIH TLV or OSHA PEL assigned to this mixture. Control of exposure to below the TLV for the ingredients (see Section 2) may not be sufficient. Minimize exposure in accordance with good hygiene practice.

Appendix D (Continued)

MATERIAL SAFETY DATA SHEET (continued):   RIMline® E-2150B

SECTION 8   SPECIAL PROTECTION INFORMATION (continued)

Respiratory protection (specify type):
If needed, use MSHA-NIOSH approved respirator for organic vapors, dusts and mists with TLV not less than 0.05 mg/m³.

Protective clothing:
Gloves determined to be impervious under the conditions of use. Depending upon conditions of use, additional protection may be required such as apron, arm covers or full body suit.

Eye protection:
Chemical tight goggles; full faceshield in addition if splashing is possible.

Other protective equipment:
Eyewash station and safety shower in work area.

SECTION 9   SPECIAL PRECAUTIONS OR OTHER COMMENTS

Precautions to be taken in handling or storing:
Prevent skin and eye contact. Observe TLV limitations. Avoid breathing vapors or aerosols.

The information herein is given in good faith
but no warranty, expressed or implied, is made.

Appendix D (Continued)

Polyurethanes
Rubicon Chemicals Inc.
Wilmington, Delaware 19897
Phone (302) 575-3000 (24 Hours)

1-800 257-5547 *Jim McBriarty*

Form No.: 5744
Date: 9/23/87

SECTION 1 NAME & HAZARD SUMMARY

Material name:

  RIMline GMR-200B   *1-504-673-6141 Tom Harbourd INDUSTRIAL HYGIENIST*

Hazard summary (as defined by OSHA Hazard Communication Standard, 29 CFR 1910.1200):
  Physical hazards: None
  Health hazards: Inhalation (TLV), irritant (skin, eye), harmful   *1-800-327-6633*
    (cardiac arrhythmias)   *Chemical MTG with*
                             on a M.D on staff

    Read the entire MSDS for a more thorough evaluation of the hazards.

SECTION 2 INGREDIENTS | % | TLV (ACGIH)
  Polyol blend | | Not listed
  Halogenated phosphate ester | | Not listed
  Tertiary amine catalyst | | Not listed
  Trichlorofluoromethane (CAS 75-69-4) | | 1000 ppm,
                                                                | | ceiling

    Ingredients not precisely identified are proprietary or nonhazardous. All ingredients
    appear on the EPA TSCA inventory. Values are not product specifications. gt =
    greater than, lt = less than, ca = approximately

SECTION 3 PHYSICAL DATA
  Boiling point: No data
  Vapor pressure (mmHg at 20°C): No data
  Vapor density (air = 1): 4.7 (trichlorofluoromethane)
  Solubility in water: No data
  pH: No data
  Specific gravity: 1.16
  % Volatile by volume: No data
  Appearance and odor: Clear orange-brown liquid

SECTION 4 FIRE AND EXPLOSION HAZARD DATA
  Flash point (and method): 420.8°F, 216°C (Cleveland open cup)
  Autoignition temp.: No data
  Flammable limits (STP): No data

  Extinguishing media:
    Water fog, foam, carbon dioxide, dry chemical, halon 1211.

  Special fire fighting protective equipment:
    Self-contained breathing apparatus with full faceplece and protective clothing.

  Unusual fire and explosion hazards:
    Heavy vapors of trichlorofluoromethane can suffocate. Highly toxic decomposition
    products.

.Appendix D (Continued)

MATERIAL SAFETY DATA SHEET (continued)   RIMline GMR-200B

SECTION 5 REACTIVITY DATA

Stability:
  Stable under normal conditions.

Incompatibility (materials to avoid):
  Oxidizing agents.

Hazardous decomposition products:
  Combustion products: Carbon dioxide, carbon monoxide, nitrogen oxides, ammonia,
  phosphorus oxides, halogen, halogen acids, possible trace amounts of carbonyl
  halide.

Hazardous polymerization:
  Will not occur.

SECTION 6 HEALTH HAZARD ASSESSMENT

General:
  No toxicity information is available on this specific preparation; this health hazard
  assessment is based on information that is available on its components.

Ingestion:
  The acute oral LD_{50} in rat is probably above 5 g/kg. Relative to other materials, a
  single dose of this product is practically nontoxic by ingestion.

Eye contact:
  This material will probably irritate human eyes following contact.

Skin contact:
  Short contact periods with human skin are not likely to result in skin irritation;
  repeated and/or prolonged contact can induce skin irritation.

Skin absorption:
  Systemically toxic concentrations will probably not be absorbed through the skin in
  man.

Inhalation:
  A rat 4-hour LC_{50} for trichlorofluoromethane is reported to be 26,400 ppm. Exposures
  to high vapor concentrations of trichlorofluoromethane induces toxicity progressing
  from giddiness, weakness, dizziness, nausea to unconsciousness. In susceptible
  individuals, cardiac sensitization to circulating epinephrine-like compounds can
  result in sudden fatal cardiac arrhythmias.

Other effects of overexposure:
  No other adverse clinical effects are known to be associated with exposures to this
  material.

First aid procedures:
  Skin: Wash material off the skin with plenty of soap and water. If redness, itching
  or a burning sensation develops, get medical attention. Wash contaminated clothing
  and decontaminate footwear before reuse.
  ——continued——

Appendix D (Continued)

MATERIAL SAFETY DATA SHEET (continued)

RIMline GMR-200B

SECTION 6   HEALTH HAZARD ASSESSMENT (continued)

First aid procedures (continued):

Eyes: Immediately flush with plenty of water. After initial flushing, remove any contact lenses and continue flushing for at least 15 minutes. Have eyes examined and treated by medical personnel.

Ingestion: Give one or two glasses of water to drink. If gastrointestinal symptoms develop, consult medical personnel. (Never give anything by mouth to an unconscious person.)

Inhalation: Remove victim to fresh air. If cough or other respiratory symptoms develop, consult medical personnel. If not breathing, give artificial respiration, preferably mouth-to-mouth. If breathing is labored, give oxygen. Consult medical personnel.

Note to Physician: Product is an asphyxiant and can induce cardiac sensitization to circulating epinephrine-like compounds resulting in potentially fatal heart arrhythmias. Do not administer adrenaline or similar sympathomimetic drugs following overexposure to this product or allow vigorous exercise for 24 hours following potentially toxic exposure.

SECTION 7   SPILL OR LEAK PROCEDURES

Steps to be taken in case material is released or spilled:

Ventilate spill area. Wear skin, eye and respiratory protection during cleanup. Soak up liquid with absorbent and shovel into waste container. Wash residue from spill area with hot water containing detergent and flush to a sewer serviced by a wastewater treatment facility.

Disposal method:

Discarded product is not a hazardous waste under RCRA, 40 CFR 261.

Container disposal:

Empty container retains product residue. Observe all hazard precautions. Do not distribute, make available, furnish or reuse empty container except for storage and shipment of original product. Remove all hazardous product residue and puncture or otherwise destroy empty container before disposal.

SECTION 8   SPECIAL PROTECTION INFORMATION

TLV or suggested control value:

No TLV assigned to this mixture. Control of exposure to below the TLV for the ingredients (see Section 2) may not be sufficient. Minimize exposure in accordance with good hygiene practice.

Ventilation:

Ventilate low-lying areas such as sumps or pits where dense vapors of trichlorofluoromethane may collect. Provide local exhaust if TLV is exceeded.

Respiratory protection (specify type):

Not normally needed if controls are adequate. If needed, use MSHA-NIOSH approved respirator for organic vapors. In high concentrations or in oxygen-deficient atmospheres, use MSHA-NIOSH approved self-contained breathing apparatus (SCBA).

Protective clothing:

Impervious gloves and apron.

Appendix D (Continued)

MATERIAL SAFETY DATA SHEET (continued)   RIMline GMR-200B

SECTION 8   SPECIAL PROTECTION INFORMATION (continued)

Eye protection:
   Chemical tight goggles; full faceshield in addition if splashing is possible.

Other protective equipment:
   Eyewash station and safety shower in work area.

SECTION 9   SPECIAL PRECAUTIONS OR OTHER COMMENTS

Precautions to be taken in handling or storing:
   Prevent skin and eye contact. Observe TLV limitations. Avoid breathing vapors or
   aerosols. Store in a cool area with good ventilation. Keep vapors away from high
   temperature surfaces to avoid toxic and corrosive decomposition products. Enforce NO
   SMOKING rules in areas of use.

The information herein is given in good faith
but no warranty, expressed or implied, is made.

Appendix D (Continued)

Polyurethanes
Rubicon Chemicals Inc.
Wilmington, Delaware 19897
Phone (302) 575-3000 (24 Hours)

Form No.: 5213h(A)
Date: 9/23/87

SECTION 1   NAME & HAZARD SUMMARY

Material name:
PRIMIine GMR-200A

Hazard summary (as defined by OSHA Hazard Communication Standard, 29 CFR 1910.1200):
Physical hazards: Unstable
Health hazards: Inhalation (TLV), irritant (eye, skin, mucous membranes, skin
   sensitizer), harmful (respiratory sensitizer, lung injury)

Read the entire MSDS for a more thorough evaluation of the hazards.

SECTION 2   INGREDIENTS

	%	TLV (ACGIH)
4,4'-Diphenylmethane-diisocyanate (MDI, CAS 101-68-8)		0.02 ppm (ceiling)
Similar structure oligomers (CAS 9016-87-9)		Not listed
Modified MDI		Not listed

Ingredients not precisely identified are proprietary or nonhazardous. All
ingredients appear on the EPA TSCA Inventory. Values are not product
specifications. gt = greater than, lt = less than, ca = approximately

---

SECTION 3   PHYSICAL DATA

Boiling point: Decomposes at 646°F, 341.1°C
Vapor pressure (mmHg at 20°C): Below 0.0001
Vapor density (air = 1): 8.6
Solubility in water: Reacts
pH: Not applicable
Specific gravity: 1.2
% Volatile by volume: Negligible
Appearance and odor: Light brown viscous liquid with slight aromatic odor

---

SECTION 4   FIRE AND EXPLOSION HAZARD DATA

Flash point (and method): 450°F, 232°C (Cleveland open cup)
Autoignition temp.: No data
Flammable limits (STP): No data

Extinguishing media:
   Dry chemical, foam, carbon dioxide, halon 1211. If water is used, use very large
   quantities. The reaction between water and hot isocyanate may be vigorous.

Special fire fighting protective equipment:
   Self-contained breathing apparatus with full facepiece and protective clothing.

Unusual fire and explosion hazards:
   Water contamination will produce carbon dioxide. Do not reseal contaminated
   containers as pressure buildup may rupture them.

Appendix D (Continued)

MATERIAL SAFETY DATA SHEET (continued)

RIMline GMR-200A

SECTION 5 REACTIVITY DATA

Stability:
Stable under normal conditions.

Incompatibility (materials to avoid):
This product will react with any materials containing active hydrogens such as water, alcohol, ammonia, amines, alkalies, and acids. The reaction with water is very slow under 50°C but is accelerated at a higher temperature and in the presence of alkalies, tertiary amines, and metal compounds. Some reactions can be violent.

Hazardous decomposition products:
Combustion products: Carbon dioxide, carbon monoxide, nitrogen oxides, traces of hydrogen cyanide.

Hazardous polymerization:
May occur. High temperatures and the presence of alkalies, tertiary amines, and metal compounds will accelerate polymerization.

SECTION 6 HEALTH HAZARD ASSESSMENT

General:
This health hazard assessment is based on information from the International Isocyanate Institute Technical Information Document #4, recommendations for the handling of 4,4'-Diphenylmethane-diisocyanate, MDI, monomeric and polymeric.

Ingestion:
The acute oral LD_{50} in rat is probably above 10 g/kg. Relative to other materials, a single dose of this product is practically nontoxic by ingestion, Hodge, H.C. and Sterner, J.H., American Industrial Hygiene Association Quarterly, 10:4, 93, Dec. 1949. Irritation of the mouth, pharynx, esophagus and stomach can develop following ingestion.

Eye contact:
This material will probably irritate human eyes following contact.

Skin contact:
No irritation is likely to develop following contact with human skin. Dermatitis and skin sensitization can develop after repeated and/or prolonged contact with human skin.

Skin absorption:
Systemically toxic concentrations are unlikely to be absorbed through the skin in man.

Inhalation:
No harmful effects occurred when rats were exposed acutely for 6-8 hours to air saturated with vapors of a similar material. However, evidence suggests that MDI can induce asthma-like respiratory sensitization that is similar to TDI sensitization. Vapors and aerosols can irritate eyes, nose and respiratory passages. May result in permanent decreases in lung function if exposure to MDI is sufficiently severe and prolonged.

Other effects of overexposure:
No other adverse clinical effects are known to be associated with exposures to this material.

Appendix D (Continued)

MATERIAL SAFETY DATA SHEET (continued)

RIMline GMR-200A

SECTION 6 HEALTH HAZARD ASSESSMENT (continued)

First aid procedures:

Skin: Wash material off the skin with copious amounts of soap and water. If redness, itching or a burning sensation develops, get medical attention.

Eyes: Immediately flush with copious amounts of water for at least 15 minutes and have eyes examined and treated by medical personnel.

Ingestion: Give one or two glasses of water to drink. If gastrointestinal symptoms develop, consult medical personnel. (Never give anything by mouth to an unconscious person.)

Inhalation: Remove victim to fresh air. If cough or other respiratory symptoms develop, consult medical personnel.

SECTION 7 SPILL OR LEAK PROCEDURES

Steps to be taken in case material is released or spilled:

Wear skin, eye and respiratory protection during cleanup. Mix with an absorbent and shovel into waste container. Cover container, but do not seal, and remove it from the work area. Prepare a decontamination solution of 0.2-0.5% liquid detergent and 3-8% concentrated ammonium hydroxide in water (5-10% sodium carbonate may be substituted for the ammonium hydroxide.) Treat the spill area with the decontamination solution, using about 10 parts of the solution for each part of the spill, and allow it to react for at least 10 minutes. Carbon dioxide will be evolved, leaving insoluble polyureas.

For transportation spills, call Chemtrec (Chemical Transportation Emergency Center), (800) 424-9300.

Disposal method:

Slowly stir the isocyanate waste into the decontamination solution described above, using 10 parts of the solution for each part of isocyanate. Let stand for 48 hours, allowing the evolved carbon dioxide to vent away. Neutralize the waste. Neither the solid nor the liquid portion is a hazardous waste under RCRA, 40 CFR 261.

Container disposal:

Drums must be decontaminated in properly ventilated areas by personnel protected from the inhalation hazards of isocyanate vapors.

1. Fill drum with decontamination solution described above, making sure all contaminated areas are in contact with the decontamination solution.
2. Leave drum soaking unsealed for 48 hours.
3. Drain and triple rinse empty container. Pour rinse solution into drain or sewer serviced by a wastewater treatment facility.
4. Puncture or otherwise destroy the rinsed container before disposal.

SECTION 8 SPECIAL PROTECTION INFORMATION

TLV or suggested control value:

ACGIH TLV and OSHA PEL for MDI is 0.02 ppm ceiling; NIOSH recommends 0.005 ppm TWA and 0.02 ppm STEL (Short Term Exposure Limit). Control limits do not apply to previously sensitized individuals. Sensitized individuals should be removed from any further exposure. The ACGIH TLV's shown in SECTION 2 are also OSHA PEL's (Permissible Exposure Limits).

Appendix D (Continued)

MATERIAL SAFETY DATA SHEET (continued)   RIMline GMR-200A

SECTION 8 SPECIAL PROTECTION INFORMATION (continued)

Ventilation:
Use local exhaust to keep exposures to a minimum.

Respiratory protection (specify type):
Because of the low vapor pressure, ventilation is usually sufficient to keep vapors below the TLV at room temperatures. Exceptions are when the material is sprayed or heated. If necessary, use a MSHA-NIOSH approved positive pressure supplied air respirator with a full face piece. For emergencies use a positive pressure self-contained breathing apparatus.

Protective clothing:
Impervious gloves. Depending upon conditions of use, additional protection may be required such as apron, arm covers or full body suit.

Eye protection:
Chemical tight goggles; full faceshield in addition if splashing is possible.

Other protective equipment:
Eyewash station in work area.

SECTION 9 SPECIAL PRECAUTIONS OR OTHER COMMENTS

Precautions to be taken in handling or storing:
Prevent skin and eye contact. Observe TLV limitations. Avoid breathing vapors or aerosols. A sensitized individual should not be exposed to the product which caused the sensitization. Store in tightly sealed containers to protect from atmospheric moisture. Store at a temperature of 60-100°F.

The information herein is given in good faith
but no warranty, expressed or implied, is made.

Appendix D (Continued)

 Polyurethanes
Rubicon Chemicals Inc.
Wilmington, Delaware 19897
Phone (302) 575-3000 (24 Hours)

003207

Form No.: 6189
Date: 03/30/87

SECTION 1 NAME & HAZARD SUMMARY

Material name:
RIMline™ R-2150A

Hazard summary (as defined by OSHA Hazard Communication Standard, 29 CFR 1910.1200):
Physical hazards: Unstable
Health hazards: Inhalation (TLV), irritant (eye, skin, mucous membranes, skin sensitizer), harmful (respiratory sensitizer, lung injury)

Read the entire MSDS for a more thorough evaluation of the hazards.

SECTION 2 INGREDIENTS

	%	TLV (ACGIH)
4,4'-Diphenylmethane-diisocyanate (MDI, CAS 101-68-8)		0.02 ppm, ceiling*
Modified MDI		Not listed

*1986 "Notice of intended change" to TLV of 0.005 ppm 8-hour TWA

Ingredients not precisely identified are proprietary or nonhazardous. All ingredients appear on the EPA TSCA Inventory. Values are not product specifications. gt = greater than, lt = less than, ca = approximately

SECTION 3 PHYSICAL DATA

Boiling point: Decomposes at 646°F, 341.1°C
Vapor pressure (mmHg at 20°C): Below 0.0001
Vapor density (air = 1): 8.6
Solubility in water: Reacts
pH: Not applicable
Specific gravity: 1.2
% Volatile by volume: Negligible
Appearance and odor: White to pale yellow liquid with faint, sweet odor

SECTION 4 FIRE AND EXPLOSION HAZARD DATA

Flash point (and method): 417°F, 214°C (Cleveland open cup)
Autoignition temp.: No data
Flammable limits (STP): No data

Extinguishing media:
Dry chemical, foam, carbon dioxide, halon 1211. If water is used, use very large quantities. The reaction between water and hot isocyanate may be vigorous.

Special fire fighting protective equipment:
Self-contained breathing apparatus with full facepiece and protective clothing.

Unusual fire and explosion hazards:
Water contamination will produce carbon dioxide. Do not reseal contaminated

Appendix D (Continued)

TERIAL SAFETY DATA SHEET (continued)   RIMline™ R-2150A

CTION 5 REACTIVITY DATA

Stability:
  Stable under normal conditions.

Incompatibility (materials to avoid):
  This product will react with any materials containing active hydrogens, such as water, alcohol, ammonia, amines, alkalies and acids. The reaction with water is very slow under 50°C, but is accelerated at higher temperatures and in the presence of alkalies, tertiary amines, and metal compounds. Some reactions can be violent.

Hazardous decomposition products:
  Combustion products: Carbon dioxide, carbon monoxide, nitrogen oxides, traces of hydrogen cyanide.

Hazardous polymerization:
  May occur. High temperatures in the presence of alkalies, tertiary amines and metal compounds will accelerate polymerization. Possible evolution of carbon dioxide gas may rupture closed containers.

CTION 6 HEALTH HAZARD ASSESSMENT

General:
  This health hazard assessment is based on information from the International Isocyanate Institute Technical Information Document No. 4, recommendations for the handling of 4,4'-Diphenylmethane-diisocyanate, MDI, monomeric and polymeric and on information that is available on a similar preparation.

Ingestion:
  The acute oral LD_{50} in rat is probably above 5 g/kg. Relative to other materials, a single dose of this product is practically nontoxic by ingestion. Irritation of the mouth, pharynx, esophagus and stomach can develop following ingestion.

Eye contact:
  This material will probably irritate human eyes following contact.

Skin contact:
  Short contact periods with human skin are not usually associated with skin irritation; repeated and/or prolonged contact can induce skin irritation. Dermatitis and skin sensitization can develop after repeated and/or prolonged contact with human skin.

Skin absorption:
  This product will probably not be absorbed through human skin.

Inhalation:
  No harmful effects occurred when rats were exposed acutely for 6-8 hours to air saturated with vapors of a similar isocyanate material. However, evidence suggests that MDI can induce an asthma-like respiratory sensitization that is similar to TDI sensitization. Vapors and aerosols can irritate eyes, nose and respiratory passages and may result in permanent decreases in lung function if exposure to MDI is sufficiently severe or prolonged.

Appendix D (Continued)

MATERIAL SAFETY DATA SHEET (continued)   RIMline™ E-2150A

SECTION 6 HEALTH HAZARD ASSESSMENT (continued)

Other effects of overexposure:
Effects associated with the use of isocyanates include respiratory and skin
sensitization, respiratory and eye irritation from vapor and liquid. Exposure to
aerosols and mists represent a greater risk of insult because aerosols represent
greater atmospheric concentrations than concentrations of vapors alone.

First aid procedures:
_____ water _____ off the skin with plenty of soap and water. If redness, itching
or a burning sensation develops, get medical attention. Wash contaminated clothing
and decontaminate footwear before reuse.
Eyes: Immediately flush with plenty of water. After initial flushing, remove any
contact lenses and continue flushing for at least 15 minutes. Have eyes examined and
treated by medical personnel.
Ingestion: Do not induce vomiting. Give one or two glasses of water to drink and
refer victim to medical personnel. (Never give anything by mouth to an unconscious
person.)
Inhalation: Remove victim to fresh air. If not breathing, give artificial
respiration, preferably mouth-to-mouth. If breathing is labored, give oxygen.
Consult medical personnel.
Note to Physician: Possible mucosal injury may contraindicate the use of gastric
lavage following ingestion.

SECTION 7 SPILL OR LEAK PROCEDURES

Steps to be taken in case material is released or spilled:
Wear skin, eye and respiratory protection during cleanup. Soak up liquid with
absorbent and shovel into waste container. Cover container, but do not seal, and
remove from work area. Prepare a decontamination solution of 0.2-5% liquid detergent
and 3-8% concentrated ammonium hydroxide in water (5-10% sodium carbonate may be
substituted for the ammonium hydroxide). Treat the spill area with the
decontamination solution, using about 10 parts of solution for each part of the spill,
and allow it to react for at least 10 minutes. Carbon dioxide will be evolved,
leaving insoluble polyureas. For major spills, call CHEMTREC (Chemical Transportation
Emergency Center) at 800-424-9300.

Disposal method:
Slowly stir the isocyanate waste into the decontamination solution described above
using 10 parts of the solution for each part of the isocyanate. Let stand for 48
hours, allowing the evolved carbon dioxide to vent away. Neutralize the waste.
Neither the solid nor the liquid portion is a hazardous waste under RCRA 40 CFR 261.

Container disposal:
Drums must be decontaminated in properly ventilated areas by personnel protected from
the inhalation hazards of isocyanate vapors.
1. Fill drum with decontamination solution described above, making sure all
   contaminated areas are in contact with the decontamination solution.
2. Leave drum soaking unsealed for 48 hours.
3. Drain and triple rinse empty container. Pour rinse solution into drain or sewer
   serviced by a wastewater treatment facility.
4. Do not distribute, make available, furnish or reuse empty container except for
   storage and shipment of original product.
5. Puncture or otherwise destroy the rinsed container before disposal.

Appendix D (Continued)

RIMline™ R-2150A

SECTION 8 SPECIAL PROTECTION INFORMATION

TLV® or suggested control value:
No ACGIH TLV or OSHA PEL assigned to this mixture. Control of exposure to below the TLV for the ingredients (see Section 2) may not be sufficient. Minimize exposure in accordance with good hygiene practice. ACGIH TLV for MDI is 0.02 ppm ceiling. NIOSH recommends 0.005 ppm TWA and 0.02 ppm STEL. In 1986, ACGIH announced that it intends to change the TLV of MDI to 0.005 ppm 8-hour TWA. These control limits do not apply to previously sensitized individuals. Sensitized individuals should be removed from any further exposure.

Ventilation:
Provide local exhaust to remove dusts or aerosols and to maintain vapor concentration below the TLV.

Respiratory protection (specify type):
Because of the lower vapor pressure, ventilation is usually sufficient to keep vapors below the TLV at room temperatures. Exceptions are when the material is sprayed or heated. If necessary, use a MSHA-NIOSH approved positive pressure supplied air respirator with a full face piece. For emergencies, use a positive pressure self-contained breathing apparatus.

Protective clothing:
Gloves determined to be impervious under the conditions of use. Depending upon conditions of use, additional protection may be required such as apron, arm covers or full body suit.

Eye protection:
Chemical tight goggles; full faceshield in addition if splashing is possible.

Other protective equipment:
Eyewash station and safety shower in work area.

SECTION 9 SPECIAL PRECAUTIONS OR OTHER COMMENTS

Precautions to be taken in handling or storing:
Prevent skin and eye contact. Observe TLV limitations. Avoid breathing vapors or aerosols. A sensitized individual should not be exposed to the product which caused the sensitization. Store in tightly sealed containers to protect from atmospheric moisture. Store at a temperature of 60-100°F. Protect from freezing.

The information herein is given in good faith
but no warranty, expressed or implied, is made.

APPENDIX E

A. Firm's Name, Address, and Phone Number

ONEROSE PLASTICS

Amerim Corporation
G-6437 Lennon Road
Swartz Creek, Michigan 48471   Phone No.: (313)635-8475

Swartz Creek, MI 48473
Any Hazard's Reported - yes ☒ no ☐
                          yes ☒ no ☐
Site Map
(for Dept. use only)

MAKES POLYURETHANE CAR PARTS

S-SYSTEM 215-D

B. Emergency Contacts for After Hours Entry
(Including Private Alarm/Security Co.)

Name/Title	Phone # For After Hours Entry
1. Robert J. Zeffero	(313)629-5610 Ex (President)
2. Robert B. Zeffero	(313)694-7196 (General Manager)
3. Leeanne Anderson	(313)629-9512 (Office Manager)
4. Mike Baur	(517)288-6036 (Supervisor)

Bill Paronski - Pres

C. Helpful Emergency Information

Self-contained breathing apparatus with full facepiece and
protective clothing should be worn per MSDS instructions.

ICI POLYURETHANES
STERLING HEIGHTS, MICH
268-9010

D. Hazardous Chemicals

Chemical Name	MSDS	Circle All That Apply	Maximum Amount lbs. or gals.	Avg. Daily Amount lbs. or gals.	Location	Physical Hazards							Health Hazards			
						Combustible Liq.	Compressed Gas	Corrosive	Explosive	Flammable	Organic Peroxide	Oxidizer	Water Reactive	Carcinogen	Highly Toxic	Irritant	Skin Hazard
Isocyanate	Yes ☒ No ☐	Pure – Solid Mix – Gas Liquid	550 gals.	225 gals.	Stored in 55 gal. drums (See Map)											(mix. db)	☒
	Yes ☐ No ☐	Pure – Solid Mix – Gas Liquid															
	Yes ☐ No ☐	Pure – Solid Mix – Gas Liquid															
	Yes ☐ No ☐	Pure – Solid Mix – Gas Liquid															
	Yes ☐ No ☐	Pure – Solid Mix – Gas Liquid															
	Yes ☐ No ☐	Pure – Solid Mix – Gas Liquid															

32

Appendix E (Continued)

Chemical Name	MSDS	Circle All That Apply	Maximum Amount lbs. or gals.	Avg. Daily Amount lbs. or gals.	Location	Physical Hazards								Health Hazards		
						Combustible Liq.	Compressed Gas	Corrosive	Explosive	Flammable	Organic Peroxide	Oxidizer	Water Reactive	Carcinogen	Highly Toxic	Irritant Skin Hazard
	Yes ☐ No ☐	Pure – Solid Mix – Gas Liquid														
	Yes ☐ No ☐	Pure – Solid Mix – Gas Liquid														
	Yes ☐ No ☐	Pure – Solid Mix – Gas Liquid														

Offices

Material Storage

Machinery

Robert B. Zeffero
Name and Official Title of Owner or
Owners Authorized Representative

Robert B. Zeffero
Signature

3/1/88
Date

Appendix E (Continued)

CHARTER TOWNSHIP OF FLINT FIRE DEPARTMENT

G-5331 REUBEN STREET
TELEPHONE - EMERGENCY: 732-4411

FLINT, MICHIGAN 48504
BUSINESS: 732-4413

The Flint Township Fire Department requests your cooperation in compiling the information necessary to implement the Michigan Right-to-Know Law. This information is necessary so that the Flint Township Fire Department and it's firefighters will have up-to-date information on the hazardous chemicals in your place of business. Also, this information will help us to better serve you and your business in the event of an emergency.

If you manufacture, use, or store hazardous chemicals (see below), this information is required under Act 80 amended P.A. 154 of 1974, the Michigan Occupational Safety and Health Act. Section 14 of P.A. 154 and Act 67 amended Act 207 P.A. of 1941, the Michigan Fire Prevention Code.

Please fill out the enclosed forms within 30 days and return them to the Flint Township Fire Department at the above address, attention Right-to-Know Division. If you do not manufacture, use, or store hazardous chemicals (see below), please complete items A, B, C, & D on page 3. Thank you in advance for your cooperation.

HAZARDOUS CHEMICAL LIST

Listed with each chemical is the recommended minimum quantity, individual, or aggregate that should be reported if located on site at any time. Individual chemicals listed below which may be in smaller quantities and present a hazard should be considered for reporting.

CHEMICAL TYPE	QUANTITY
Poison A	Any Quantity
Flammable Gas	100 gal. water capacity
Non-Flammable Gas	100 gal. water capacity
Poison B	50 lbs.
Flammable Liquid	100 gal.
Combustible Liquid	1,000 gal.
Corrosives - Liquid Solid	100 gal., 50 lbs.
Irritating Material - Liquid Solid	100 gal., 50 lbs.
Explosives & Blasting Agents	
(Not Including Class C Explosives)	Any Quantity
Radioactive Material (Yellow III Label)	Any Quantity
Flammable Solid (Dangerous When Wet)	50 lbs.
Spontaneously Combustible Material	50 lbs.
Oxidizer	50 lbs.
Organic Peroxide	Any Quantity
Carcinogens	Any Quantity
Biohazardous Material	Any Quantity
Etiologic Agents	Any Quantity
EPA "Extremely Hazardous Substances	Any Quantity
Acutely Hazardous Waste	2.2 lbs.
Other Hazardous Waste	220 lbs:
Other Hazardous Chemicals Not Otherwise Identified	10,000 lbs.

Dedicated to Promoting Safety, Saving Lives, Fighting Fire

Appendix E (Continued)

GUIDELINES & INSTRUCTIONS FOR COMPLETING SURVEY FORMS

Following are suggestions for completing the Hazardous Chemical Survey Form (page 3). Please type or print with a soft lead pencil or felt tip pen to aid in reproduction for our emergency response files.

A. Firm's Name, Address, and Phone Number.

B. Phone numbers where we can reach someone for help in the event of an emergency during off hours.

C. For those who don't manufacture, use, or store hazardous chemicals (on front page). Any information regarding your business you feel could be helpful during an emergency, please put in this section.

D. Hazardous chemicals (as defined in Federal Hazard Communication Survey (on front page).

   1. Enter the chemical name or common name of each hazardous chemical.

   2. If you have Material Safety Data Sheets or Product Bulletins on the chemicals you've listed, please indicate by checking yes or no under MSDS.

   3. Circle ALL applicable descriptors: pure or mix, and solid liquid or gas.

   4. Maximum Amount - For each hazardous chemical, estimate the greatest amount present at your facility on any day during the reporting period. Enter the amount and the measuring standard gallons or pounds.

   5. For each hazardous chemical estimate the average or typical amount in gallons or pounds that could be found in your facility on an average day.

   6. Physical and Health Hazards. For each chemical you have listed, check all the physical and health hazards boxes that apply. These hazard categories are defined in the OSHA Hazard Communication Stand 29 CFR 1910.1200.

   7. Location - A brief description of the location of each hazardous chemical, how stored (bags, drums, above or below ground storage, etc.) or anything else you feel may help us to handle an emergency involving this material.

E. A site map to see where the chemicals are relative to other site features. This does not need to be a detailed drawing, a freehand sketch showing relative distances, storage areas, shut-offs, access, etc., would be fine.

F. A copy of your firm's contingency plan (if available).

G. We would like to know who supplied the information on this form so that if questions arise we have someone to contact.

NOTE: If there is any other information that you think would be helpful to us, please submit on a separate sheet. If you have any questions, contact Firefighter Chesnut at 732-4413. Thank you for taking the time to fill out this form.

Sincerely,

D. S. Rowley

Chief Donald S. Rowley

APPENDIX F

GENESEE COUNTY FIRE COORDINATORS

SCOPE:

Coordinators are a tool to assist a fire department at a major incident with the movement of equipment, personnel, and/or other problems.

Coordinators will, at the discretion of the officer in-charge of the incident, take whatever steps necessary to fill the request of that officer. Coordinators shall assess the overall situation and initiate actions (move-ups) required to provide adequate protection to all areas of the county.

HOW ACTIVATED:

The officer in-charge of the incident can activate the coordinators at his/her discretion by contacting Flint Township Fire Department (27) by radio or telephone and request the coordinators be activated.

When Flint Township Fire is notified, they will activate the county master en-coder and request the coordinators to be activated. If the incident is in the north county, the south county coordinators will be activated first. If the incident is in the south county, the north county coordinators will be activated first. The activation of the coordinators is for information only, and does not necessitate the staffing of stations, unless requested to do so.

When county coordinators are activated, Flint Township Fire will call the Amateur Radio Operators, One Amateur Radio Operator will report to Flint Township Fire and one will go to the command post at scene as alternate communications.

COORDINATORS OPERATIONS:

The request of the coordinators shall be considered same as a request from the officer in-charge, under the terms of the Mutual Aid Pact.

All on-scene radio communication will be conducted thru the coordinators, not thru base radios, once coordinators are in operation.

All apparatus moving up shall be under the control of the coordinators and radio communications shall be directed to the coordinators. Base radios are not to communicate with apparatus unless operating another incident within their own fire district.

Incidents using coordination shall be handled on the Mutual Aid Radio Frequency for all units at the scene. The coordinators will initiate this if necessary.

Units going to the incident shall change to the Mutual Aid Frequency when they are within two minutes of the incident.

During times of coordination, all radios not involved in the incident shall remain on their normal operating frequency.

The coordinators shall remain active as necessary until the status of equipment is at a normal and adequate level.

Revised Sept. 1988

FLINT TOWNSHIP
BASIC DISASTER PLAN

Township Supervisor

Director, Genesee County
Office of Civil Defense

FLINT TOWNSHIP
BASIC DISASTER PLAN

INTRODUCTION

The Flint Township Basic Disaster Plan has been developed to insure the residents of Flint Township that local government will be able to respond to major disasters in a prompt, orderly, and organized fashion. Experiences of the past have shown that a well thought out disaster plan can facilitate Government's response to disasters. Communities with disaster plans have been shown to be more capable of responding to disasters than those without them.

The Flint Township Basic Disaster Plan is designed to deal with both man-made and natural disasters. It deals most specifically with the probable results of nuclear attack. It does so for two reasons. First, it fulfills the requirements for nuclear planning as established by both the Federal and State governments. Secondly, by planning for the worst possible disaster, the plan incorporates all the possible tasks government faces in any of the smaller disasters.

The plan assigns general areas of disaster related responsibilities to given departments or individuals. In some cases it gives rather specific detailed assignments, but in most cases it simply provides a set of guidelines for the development of detailed standard operating procedures. The departments of individuals charged with performing the actual work are thus able to adapt their own day-to-day operations to disaster situations. This is in keeping with the planning precept that disaster related tasks should be assigned to those who do similar work on a day-to-day basis.

Individuals who have suggestions or comments related to this plan or the standard operating procedures manual should direct them to:

> Galen Jamison
> Flint Township Supervisor
> 1490 S. Dye Road
> Flint, MI 48532
>
> or:
>
> John H. West, Jr.
> Genesee County Office of
> Emergency Preparedness
> 1101 Beach Street
> Flint, MI 48502

BASIC PLAN INDEX

I.   LEGAL AUTHORITY

A.   Authority for this plan is provided by the Michigan Emergency Preparedness Act, Act 390, Public Acts of 1976, Section 10, as amended.

B.   Authority for this plan is established by Resolution Number 77-259 of the Genesee County Board of Commissioners to establish an emergency preparedness policy and organization dated June 14, 1977.

C.   Authority for this plan is the result of a resolution adopted by the Board of Trustees of Flint Township.

II.   HAZARDS/PURPOSE

A.   Hazards

1.   *Enemy Capabilities*

It is assumed that potential enemies of the United States have the capability of launching an attack on the United States with sufficient weapons, nuclear, biological, and chemical, to strike a high proportion of our military, industrial, and population targets at a time of their choosing. For Crisis Relocation Planning purposes, Flint Township is designated a risk area.

2.   *Natural Disasters*

Flint Township is subject to tornadoes, floods, fires of major proportions, explosions and other catastrophes occurring from natural and accidental causes.

3.   *Other Incidents*

It is also recognized that Office of Emergency Preparedness services may be needed in the event of riot or civil disturbances and industrial or nuclear accidents.

B.   Purpose

The mission of the Flint Township Basic Disaster Plan is to organize, coordinate, and direct the actions of the Township emergency services and the public, to execute prepared plans of operations in the event of enemy-caused emergency, natural disaster, catastrophic accident, or civil disturbances; to save the maximum number of lives; minimize damage to property; to receive and disseminate the attack warning; to direct the public to the best available shelter in a fallout situation; to maintain the continuity of government; to preserve vital records; and to provide support and assistance to other jurisdictions.

III.   DEFINITIONS

For the purpose of this Emergency Operations Plan, certain words and phrases used herein are defined as follows:

A.   "Disaster" means an occurrence or imminent threat of widespread or severe damage, injury, loss of life or property resulting from any natural or man-made cause, including but not limited to fire, tornado, flood, snow, ice, or wind storm, wave action, oil spills, water contamination requiring emergency action to avert danger or damage utility failure, hazardous peacetime radiological accident, major transportation accident, epidemic, air contamination, blight, drought, infestation, explosion, riot, or hostile military or paramilitary action. Riots and other civil disorders are not within the meaning of this term unless they directly result from and are an aggravating element of the disaster.

B.   "Emergency Preparedness Coordinator" shall mean the person duly appointed to coordinate emergency planning and services within Genesee County to protect the public health, safety, and welfare during emergency situations and disasters. In the absence of an appointed person, "Coordinator" shall mean the Chairperson of the Genesee County Board of Commissioners.

C.   "Local Civil Defense Liaison" shall mean Galen Jamison, Flint Township Supervisor.

D.   "Emergency Operating Center" (EOC) shall mean the Federally approved EOC located in the lower level of the Genesee County Administration Building, 1101 Beach Street, Flint, Michigan. This EOC will be the facility where heads of departments and agencies, or their representatives, will be assembled during disasters to facilitate coordinated disaster response and recovery. A secondary facility may be designated by the Coordinator or the Chairperson of the Board of Commissioners as the EOC to be used for the above purposes.

E.   "Local Disaster Control Center" shall mean the command center established by the Flint Township Disaster Plan in Section V.D.

F.   "Emergency Operations Plan" means the Genesee County Emergency Operations Plan which has been developed to assign various disaster tasks among departments and agencies and to coordinate disaster response and recovery within Genesee County. The EOC is also called the Emergency Preparedness Plan for the purposes of Act 390, Public Acts of 1976, Section 19.

G.   "Major Disaster" or "Emergency Declaration" means the designation by the President under the provisions of Public Law 93-288, upon the request of the Governor, that a disaster of major proportions has occurred and that Federal disaster assistance will be provided subject to conditions established by the President.

H.   "State of Disaster" means a declaration of executive order or proclamation by the Governor under the provisions of Act 390, Public Acts of 1976, which activates the disaster response and recovery aspects of State, county, local, and interjurisdictional disaster emergency plans and authorizes the deployment and use of any forces to which the plan or plans apply.

I.   "State of Emergency" means a declaration by the Chairperson of the Board of Commissioners which activates the disaster response and recovery aspects of the Genesee County Emergency Operations Plan and authorizes the deployment and use of any disaster relief forces to which the plan applies.

J.   "Disaster Assistance Center" shall refer to federally selected site through which all services to individuals will be channeled. It allows victims to obtain all available services in a "one-stop" fashion.

IV. KEY ASSUMPTIONS – CONCEPT OF OPERATIONS

A.   Initial and primary response to disasters is the responsibility of local government. When it has been determined that the disaster is beyond the capability of local government or where special equipment or resources are necessary to alleviate the effects of the disaster, assistance from the county may be requested. If the disaster is beyond the county's capability or requires special resources, the county may request assistance from the State government. If the disaster is beyond the State's capability or requires special resources, the State may request assistance from the Federal government.

B. It is a basic concept that emergency operations will make use of all available resources (governmental and private) to combat the effects of a disaster. Since the normal functions and organization of local government will be the primary resource around which a disaster operations organization will be developed, appropriate emergency functions are assigned to the various departments of Flint Township government in-line with normal day-to-day responsibilities, as much as possible.

C. Heads of departments assigned responsibility for emergency functions, or their representatives, will be assembled in a central facility called the EOC under the direction of the Chairperson of the Genesee County Board of Commissioners or at the Disaster Control Center (DCC) under the Supervisor. The Genesee County Emergency Preparedness Coordinator will function as the coordinator of all activities within the EOC. The Civil Defense Liaison Officer will serve as the coordinator at all activities in the local DCC. The EOC or DCC will have a communications capability so that field elements of all disaster relief forces can be directed and controlled by the appropriate EOC or DCC staff and so that required information may be received, recorded, plotted, analyzed, and so that timely decisions may be in response to a disaster. Common information will be displayed for all department heads to see and use. Communications will be established with the State, local communities and agencies, and adjacent jurisdictions.

D. The Genesee County Civil Defense Disaster Plan will be used to supplement Flint Township's ability to respond to disasters. In turn the resources of Flint Township are made available to the County Plan as needed and available.

V. EXECUTION/ACTIVATION

A. Activation of this plan will be initiated by the declaration of a state of emergency by the Supervisor or by the Chairperson of the Genesee County Board of Commissioners upon the request of Flint Township.

B. Whenever the Chairperson deems a disaster is beyond the control of local, public or private agencies and that county, State, Federal, or military assistance may be required, he may act for the Township and request the Genesee County Emergency Preparedness Coordinator to request that the Governor declare a state of disaster.

Such request will be submitted by the coordinator to the Michigan State Police District Emergency Services Coordinator in accordance with Section 14, Act 390, Public Acts of 1976. The Chairperson shall convene the Board of Commissioners as soon as practical for their affirmative action. The Supervisor will convene the Board of Trustees as soon as practical for their affirmative action in all cases.

C. Upon the declaration of a state of disaster by the Governor, this plan will be automatically activated, if not activated previously by the Supervisor or the Chairperson of the County Board of Commissioners.

D. The Supervisor may exercise all emergency power and authority as specified herein. Whenever a situation requires, or is likely to require, that the Supervisor invoke such power and authority, he shall, as soon as reasonably expedient, convene the Board of Trustees to perform its legislative and administrative duties as the situation demands, and shall report to that body relative to emergency activities.

The Local Civil Defense Liaison will be responsible for identifying and staffing the local Disaster Control Center, as authorized by the Supervisor. The Civil Defense Liaison Officer will be responsible for coordinating the efforts of county, State, and Federal agencies as required. The Assistant Civil Defense Liaison shall serve as the Supervisor's chief advisor on dealing with the emergency or disaster.

The DCC for Flint Township is located at the Flint Township Fire Department, 5331 Reuben St., Flint, Michigan.

Secondary, 3327 Flushing Road.

It is also recognized that the delivery of State and Federal individual assistance program may require the establishment of a Disaster Assistance Center to facilitate the delivery of such services. All agencies providing such services will be represented in the Disaster Assistance Center. Suggested locations for Flint Township are:

1. Dye Elementary
   1174 Graham Rd.
   Flint, MI 48532

2. Senior Citizens Center
   2071 S. Graham Rd.
   Flint, MI 48532

3. Carman High School
   1300 N. Linden
   Flint, MI 48532

It is also recognized that a local coordinating officer for Flint Township may be required to work with Federal officials. The Supervisor, Mr. Galen Jamison, or assistant civil defense liaison will serve in this capacity.

Warning

Nuclear Attack:

The National Warning System (NAWAS) is used to distribute nuclear attack related information. The NAWAS warning point for the Flint area is the Michigan State Police Post #35. The post will provide warning of relocation, impending attack or natural disaster to key agencies. The city of Flint Police Department is designated as the secondary warning point for the Flint Township area.

The Emergency Broadcast System will also be utilized to disseminate such warning.

Natural Disasters

The National Atmospheric and Oceanic Administration have established a weather broadcasting system which disseminates warnings on a twenty-four hour basis. The NOAA Weather Radio is the fastest means of obtaining weather alerts, watches and warnings. Flint Township has a weather alert radio located at the Flint Township Fire Department, 5331 Reuben, Flint, Michigan.

The broadcast media serving the Flint Township area also cooperates in disseminating information to the public.

Other Warning Systems

Communications:

Telephones:

Flint Township has a telephone system capable of handling 14 incoming calls at any one time. They are located at the Flint Township Hall, 1490 South Dye Road, Flint, Michigan. The Flint Township Fire Department has the capability of handling 4 incoming calls.

Police:

The Flint Township Police Department operates a radio system on the following frequencies:

	Trans.	Rec.
Lein	156.030	155.610
Central	155.850	155.445
Car to Car	155.880	155.880
Emergency Car to Car	155.865	155.865

It has eight (8) mobile radios and eleven (11) portable radios. Its base station is located at the Flint Township Police Department, 1490 S. Dye Road, Flint, Michigan.

Fire:

The Flint Township Fire Department operates a radio system on the following frequencies:

1. South End Normal Dispatch Channel	154.190
2. Mutual Aid Dispatch Channel	154.280
3. North End Dispatch Channel	154.145
4. State-wide Mutual Aid Channel	154.129

It has 12 mobile units and 18 portable units. Its 3 base stations are located at the Flint Township Fire Department, 5331 Reuben St., Flint, Michigan and at Station #2, 3327 Flushing Road, Flint, Michigan, and Station #3 at 2511 W. Bristol Road, Flint, Michigan.

Volunteer Services

Amateur Radio Emergency Services (ARES):

The ARES is a local group of ham radio operators who are available for disaster related work. They may be activated by calling (313) 767-7048.

Community Radio Watch:

The Community Radio Watch (CRW) is a police approved and trained group of Citizens Band Radio Operators. They may be activated by calling (313) 766-7444 at the Flint Police Department.

Damage Assessment

The amount and types of damage resulting from disasters has a direct bearing on the nature of the response to such disasters. Consequently, it is paramount that an accurate assessment of such damage take place immediately. Flint Township has designated the Township Assessor as its Damage Assessment Officer (DAO).

The DAO will prepare a summary of any disaster damage occurring in Flint Township. The DAO will also ensure that the County Emergency Preparedness Director is informed as well.

Should the damage be extensive enough to require State or Federal aid, Form E.S. 2, will be used to summarize the damage.

Emergency Public Information

The Genesee County Office of Civil Defense shall provide an Emergency Public Information Office (EPIO) to aid the Supervisor in dealing with the media during disasters.

The EPIO will be responsible for authenticating all information released to the public. The EPIO will also be responsible for rumor control.

The Supervisor may designate an alternate EPIO if he finds it necessary or desirable.

Law Enforcement

The Flint Township Police Chief will serve as Law Enforcement Coordinator for Flint Township in the event of a large scale disaster. He (or his designee) shall report to the DCC as soon as it is established.

The chief will be responsible for the protection of property and the maintenance of law and order. The chief will be responsible for securing the site of any disaster damage.

Additional aid will be sought from the Sheriff's Department and the Michigan State Police as needed. If additional aid beyond local police resources is needed, the chief may, through the Genesee County Office of Civil Defense, request that the National Guard be activated.

Fire Services

The Flint Township Fire Chief (or his designee) shall serve as the Fire Coordinator for Flint Township. The fire department will be responsible for fighting fires, coordinating rescue operations and handling chemical or other Hazmat problems including radiological monitoring and radiological defense.

The fire chief will assign a fire representative to the DCC.

The fire chief will also be responsible for soliciting additional aid under the Mutual Aid Pact for Genesee County Fire Departments.

Health and Medical Services

The Joint Hospital Disaster Preparedness Committee Medical Disaster Plan will be used to deal with large scale medical emergencies. The Health Annex to the Genesee County Basic Disaster Plan will be used to deal with large scale health threats.

Shelter

The Red Cross will be responsible for sheltering disaster victims left without housing. The Red Cross will maintain a list of available natural disaster shelters for use as temporary mass housing.

The Red Cross will also work closely with the Genesee County Office of Civil Defense and Salvation Army to relocate disaster victims in more permanent and private housing.

Should a national disaster be declared, the Red Cross will work through the Genesee County Office of Civil Defense to establish linkage with Federal housing personnel.

The Civil Defense Liaison will work with the County Office of Civil Defense to identify and mark fallout shelters in Flint Township for purposes of nuclear attack or peace time nuclear accidents.

Public Works/Debris Clearance

The Genesee County Road Commission will be responsible for assuring that the streets of Flint Township are cleared after any disaster.

The Flint Township Health Board will be responsible for debris clearance on private property when viewed as a health hazard.

The County Road Commission and private contractors within Flint Township will aid fire personnel with heavy rescue equipment as needed.

The Flint Township Building Inspector will be responsible for inspecting and labeling all disaster damage buildings.

Such labeling will be designed to protect occupants and others from health and safety hazards.

Each of the above agencies will send representatives to the DCC to assure good response coordination. Each of the above agencies will also provide the Damage Assessment Officer with damage information as appropriate.

Local Civil Defense Liaison

The Civil Defense Liaison will be responsible for:

1 Identifying a Disaster Control Center for an Alternate Disaster Control Center.

2. Identifying a DCC for an Alternate DCC.

3. Identifying two potential sites for Disaster Assistance Centers should they be needed.

4. Staffing and equipping the DCC, training and coordinating, the DCC staff.

Lines of Succession

The line of succession for the continuity of administration and government in disasters, as provided by the resolution adopted on is as follows:

1. *Chief Executive Officer*
   Galen Jamison, Township Supervisor

   1st Successor
   Township Clerk

   2nd Successor
   Township Treasurer

2. *Civil Defense Liaison*
   Township Supervisor

   Assistant Civil Defense Liaison
   Fire Chief